Raindrops From Heaven

The God Gap, Volume 3

Robert Tennant-Ralphs

Published by Beacons of Light Books and Films, 2025.

RAINDROPS FROM HEAVEN

First edition. January 27, 2025.

ISBN: 978-1068304040

Written by Robert Tennant-Ralphs.

Table of Contents

To all the friends of Bill and Bob who helped me on my path of enlightenment – many of whom I have not met.

To Dan, a wonderful, inspiring son who always knew I wasn't born yesterday.

To Svitlana Kozmin who encouraged me to write this book, Kathleen Price and Simon Wharmby who read it and guided me, and Duncan Deacon who did the clever technical stuff.

To my family who stood by me through my ups and downs.

To those who dream boldly, persevere through challenges, and believe in the transformative power of love and nature—this book is for you. May it inspire, uplift, and remind you of the limitless possibilities within.

Raindrops from Heaven -

12 Lessons of Water for Humans

By Robert Tennant-Ralphs

RAINDROPS FROM HEAVEN - Lessons of Water explores the profound ways in which water shapes our world and offers wisdom for how humanity can thrive in harmony with nature.

From being essential to humans and every living thing, to sculpting majestic geological formations to nurturing vibrant ecosystems, from serving as a vital transportation medium to providing habitats for countless species, water's roles are as diverse as they are essential. Yet, it is a resource often overlooked or misunderstood.

Through thought-provoking insights, this book reveals the remarkable lessons water can teach humans about resilience, adaptability, and interconnectedness. By the time you reach the final page, you will never see water - or life itself - in the same way again.

TABLE OF CONTENTS

Dedication

To all the friends of Bill and Bob who helped me on my path of enlightenment – many of whom I have not met.

To Dan, a wonderful, inspiring son who always knew I wasn't born yesterday.

To Svitlana Kozmin who encouraged me to write this book, Kathleen Price and Simon Wharmby who read it and guided me, and Duncan Deacon who did the clever technical stuff.

To my family who stood by me through my ups and downs.

To those who dream boldly, persevere through challenges, and believe in the transformative power of love and nature—this book is for you. May it inspire, uplift, and remind you of the limitless possibilities within.

Chapter 1 Heavenly Raindrops

I was twelve years old when I had the first of two transformative experiences that set me on a path of discovery and self-awareness—both inspired by the wonders of water.

It began as a simple homework assignment from our form teacher. She asked us to create a list of things we liked and disliked about our lives, explaining that the exercise was meant to help us focus on what brought us joy and was truly beneficial. The next day, we were to share our lists with the class.

But I was a rebel, determined to go against the grain.

That evening, I borrowed my father's laptop and began my "homework" with a twist. Instead of following the assignment's intent, I decided to write an essay about something I despised. As I reflected on the weekly Friday night baths I dreaded and the endless rain that soaked the Welsh valley where I lived, one thing stood out: I hated water.

Yet, as I delved deeper into my research, I was amazed by what I found. Water, I realised, was not just something to tolerate; it was extraordinary. Its countless, diverse functions—sustaining life, shaping our world, and even enabling the smallest of miracles—challenged my perspective. That night, my essay took an unexpected turn: from declaring my dislike for water to beginning to understand why I should appreciate it.

Later in life, I came to appreciate not just waters utility but also its beauty and miraculous qualities. That night, however, marked the beginning of my realisation that water was not something to hate, but something to admire—and perhaps even love.

I BEGAN MY ESSAY BY imagining myself as a single drop of rain, one of what I later came to call *God's teardrops*, which I will explain later. I pondered the life

I might have lived and the future that lay ahead. Always, I would be water - a particle of H_2O. That's all I could ever be, no matter my form.

I might appear as rain, ice, vapor, dew, frost, fog, hail, snow, a river, waterfall, sea, or ocean. I might also be part of milk, blood, vegetables, fruit, tea, coffee, beer, or cola. I could even reside in ice cream or any living plant or creature. No matter where I existed, I would always be water.

In one moment, I could be steam escaping from a kettle or vapor rising from a thermal spring. Days later, I might fall as rain into a valley, quenching the thirst of a bird or animal drinking from a river winding down the mountain to a distant lake or ocean. I might become saliva in an animal's mouth, seep into the soil to nourish a tree, flower or vegetable's roots, or partake in countless other functions vital to life on Earth. I am also the main component in fruit and vegetables.

Fruit is over 50% water

Vegetables are over 50% water

MY NEVER-ENDING STORY of goodness does not stop with being the main ingredient for life on Earth. Not long ago, I was in the sacred Ganges River during the Hindu pilgrimage of Kumbh Mela. Before that, I had been in the holy water used to baptise a child.

River Ganges pilgrimage

Baby Baptism

AS A DROP OF WATER, I am essential to all living things. From the perspective of creation, I am always where I am needed. Yet, despite my vital role, I have no control over my destiny. Everything that happens to me - my transformations, my journeys - unfolds without my influence.

What's more astonishing is my power to shape the world. I carve mountains, mould the Earth's surface, and alter landscapes, feats that seem unimaginable for anyone else. And yet, even in these extraordinary acts, I am not in control.

These reflections left me with profound questions:

1. Who or what sculpts our planet into its present form?
2. Who or what created the water cycle?
3. Who or what governs water's perfect distribution across the globe?
4. Who or what imbued water with the properties essential for all flora and fauna to thrive?
5. Could it all be a product of chance?

As a drop of water, I have existed since time immemorial and will continue to exist for eternity. I play a vital role in the evolution of life, yet I cannot fathom my origins.

I may never uncover definitive answers to these questions, but one truth is clear: life as we know it could not exist without water. The planet's physical form would be unrecognizable without my countless, seemingly lifeless companions.

I marvel at how water contains an invisible life-giving force, one that science acknowledges but cannot fully explain. It led me to wonder: how did every form of life - plants, insects, animals, birds, fish, and reptiles – come to rely on water for survival? What is the essence of its invisible life-giving properties?

Most forms of food derive from something that was once alive, but water, air, and soil are exceptions. Yet, all three are essential to life. This raised a question in my mind:

Which came first, the life-giving elements in water, air, and soil, or the need for them?

Logic suggests that the elements must have preceded life since early organisms could not have survived without them. Yet, how is it that life depends so profoundly on seemingly 'lifeless' invisible substances?

I realised that many other essential forces - gravity, electromagnetism, nuclear energy, light, heat, wind, and radio waves - are also invisible. If we trust their existence based on scientific evidence, should we not also consider the possibility of an invisible, *creative Intelligence* orchestrating these forces?

The harmony and precision with which these forces have supported life across billions of years defy human comprehension. Could it all exist without a *Mastermind* guiding it?

Acknowledging the existence of invisible forces led me to accept the possibility of an unseen Intelligence behind it all - a conductor of this universal symphony. For me, this understanding bridged the gap between science and spirituality. It offered a way to reconcile my agnostic past with the overwhelming evidence of a purposeful design in the universe.

This newfound belief made me see that the Creator must be as invisible as the forces they govern - an Intelligence far beyond human understanding. Could it be that those who believe in animism are closer to the truth than I thought?

Water's other functions remind us of its endless versatility. It serves as a home for fish and aquatic plants, a means of transportation. It carves landscapes, generates energy, and extinguishes fires. So, despite its *lifelessness*, water is indispensable to all life.

MY SECOND TRANSFORMATIVE experience with water occurred during a visit to Venice. This city, celebrated for its breathtaking beauty and intricate waterways, is built upon a lagoon and interwoven with canals. Yet, for me, Venice holds a much deeper meaning.

In 1980, knowing I was a chronic alcoholic, I travelled to Venice with the daughter of a doctor, determined to stop drinking. I failed, and my tears of despair only added to the countless drops of water in the canal outside the church where I wept.

Six years later, I returned to Venice alone for a weekend. By then, I had been sober for over a year. As my plane descended toward Marco Polo Airport, I opened a travel guide from my previous trip. A postcard slipped out, and what happened next would change my life forever.

The postcard depicted the Church of Santa Maria dei Miracoli. The following day, I visited it, and as I saw the very steps where I had once cried in despair, I felt an overwhelming sense of realisation: my cry for help had been heard, and I had been saved.

Since then, I have made a pilgrimage to Venice every two years. Each time, as my water bus approaches St. Mark's Square, I am reminded of how the city's extraordinary fusion of water and architecture makes it one of the world's great natural and man-made wonders. For me, those drops of water have transformed into *God's teardrops.*

Santa Maria dei Miracoli, Venice

St Mark's, Venice

Chapter 2 Which Came First?

Which came first: water or the planet? How could such perfect harmony exist between two lifeless entities without an *Intelligence* to unite them?

These questions continue to echo within me, but one conclusion remains: water's presence is a testament to an extraordinary force at work- a force that ensures life's existence and perpetuates its continuity.

This led me to new reflections. Scientists suggest that life on Earth originates from water, which may explain why every living being - plants, insects, animals, fish, birds, and reptiles - is dependent on it for survival. This made me wonder: what is the life-giving property in water that makes it so essential?

Most food, apart from certain minerals, originates from something that was once alive. But this is not true for water, air, and soil - yet all three are indispensable for life. This realisation sparked deeper questions:

- Which came first, the eater or the eaten?

- Which came first, the life-giving properties in water, air, and soil or the need for them?

It must have been the life-giving properties, for the earliest forms of life could not have existed without them. But how is it that something as *'lifeless'* as water, air, or soil can sustain so much that is *alive*? Even more perplexing is that their *life-giving* qualities are invisible!

I had already noticed that other essential forces – electro-magnetism, gravity, nuclear energy, electricity, radio waves, light, heat, and wind are also invisible. Trusting the scientific evidence for these forces, I was struck by how they orchestrate Earth's physical and geological needs with extraordinary precision and timing, enabling life to form and flourish.

How could such disparate forces work so perfectly, across billions of years, unless guided by an *all-knowing, all-creative Mastermind?*

While I couldn't fully comprehend these mighty, invisible forces, I couldn't deny their existence. This compelled me to consider the possibility of an *invisible creative Intelligence* at work. Accepting this idea helped me recognise that such Intelligence must have knowledge of all these forces, their roles, and how to orchestrate them in harmony. For a former agnostic, this was a staggering realisation - but I could think of no other explanation. Can you?

After reaching this conclusion, it seemed logical that these forces must be eternal, for something cannot come from nothing. This reinforced my belief that life has purpose, that my role - and everyone else's - has meaning.

If the fundamental forces of the universe are invisible yet capable of operating in harmony, it stands to reason that the controlling Intelligence must also be invisible. This Intelligence, infinitely greater than I could ever imagine, is in charge. It brought to mind animistic beliefs - how right they now seemed!

Water's functions extend beyond sustaining life. It serves as a mode of transport for ships, provides habitats for plants, fish, and animals, and shapes ecosystems through soil erosion and the carving of riverbanks. It generates energy as steam or hydroelectric power, helps extinguish fires, and boils kettles. In short, water's usefulness is vast and indispensable to life on Earth.

Yet water itself could not have evolved like plants, animals, or fish. So, who or what created and distributed it so perfectly, granting it both direct and indirect life-giving properties?

- Which came first: water or the planet?
- How could such a flawless relationship between two lifeless entities exist without a *Higher Power* to orchestrate it.

These questions multiplied. Here was a substance critical to life, available in sufficient abundance to perform its functions flawlessly. (A later chapter addresses water shortages in drought-prone areas.)

The answers to evolutionary questions I encountered in school and science books often credited 'the forces of nature.' I now saw these as manifestations of 'Intelligence' or 'God.' This was *the God of my understanding* - not tied to any single religion or doctrine.

For the first time, I grasped the concept of a conductor guiding the universe's intricate diversity and orchestrating the needs of its multitude of manifestations. If I could reach this understanding, surely others could too.

As a child, I had been told of an 'Eternal Father' - an unseen, untouchable, unheard presence. I had dismissed this notion and resisted those who insisted I believe it. But now, this truth had become the most profound discovery of my life, and I felt an obligation to share it. The challenge was finding the best way to do so.

※

BY NOW I WONDERED IF water relied on external *Intelligence* to guide it to where it is needed. So, I returned to imagining myself as a raindrop to find out.

I have fallen onto a pond on the outskirts of Casablanca in Morocco, Northwest Africa. Before this, I was an ordinary drop of rain like any other and now I want the world to know my story.

I am identical to every other raindrop that fell on this pond, however, I have the most amazing tale to tell.

Raindrops from Heaven

IT BECOMES VERY HOT, and the sun is beating down as usual for that time of day. Within a short time, I get so heated up that I just had to 'explode', so I evaporate. I turn into minute particles of water vapour and rise; I am lifted by a thermal current. Before long I start to cool and stop rising.

(By the way, I just defied the law of gravity: Can you do that?)

During this time, I am joined by billions of other particles of water. We bind together and become part of a cloud. I get carried along by a strong wind, which is completely outside my control, much like everything else in my life.

This makes me wonder, how does the binding come about? Is it caused by a force from above or below? And, what about the creative force behind the wind, where does the Intelligence behind that come from?

I stay in this cloud until I see far below a huge blue mass, the Mediterranean Sea. I recall that years ago I was in it when a fish swallowed me, and then, after passing through its body, I was deposited back into the sea.

As I drift over it now, I am gradually lifted higher within the cloud. It becomes very dark, and I begin to feel colder.

I travel over land and soon see below white-capped mountains, the Alps. By now the clouds are dense and I am very cold. Other particles join me, and I turn into ice. Then I fell as a snowflake along with billions of other snowflakes. Once I settle as snow on the ground, I get covered by more snow. I freeze and here I remain, quite contentedly, until the following spring.

In early April I melt in the sun and become water again. (I prefer being liquid water as life is more interesting, though I quite enjoyed the rest!)

I am on the move again. Once I had melted, I seeped into the soil and joined other particles of water that would eventually trickle out of the surface of the ground. But before that I had not lain dormant as humans may think; I had vital functions to perform. As Leonardo de Vinci said, 'water is the driving force of all nature', and nowhere is it more obvious than in soil.

With soil covering an estimated 10% of the Earth's surface, because of the ubiquitous nature of me and my siblings, we are able to perform critically important functions, *seemingly unaided*, that are essential to all that lives above ground.

Every plant, insect, bird, reptile, amphibian, and animal benefits from what we do. We are essential to every ecosystem and human habitation. With the aid of gravity, our abundance of connected pores in soil allows us to feed all life and to create springs, wells and watering holes for animals.

Although I know only a little, if this did not take foresight and need control at all times, I would eat my hat if I was human.

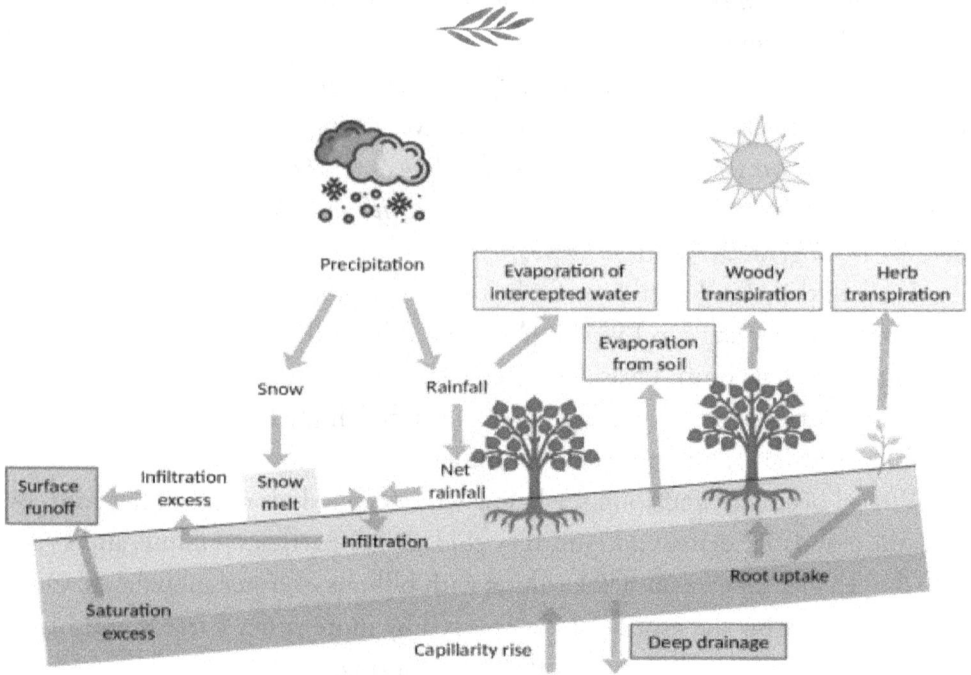

Water's interactions with the environment

Spring water Australia

Well water in India

African watering hole

ONCE MELTED, I AM ON the move again. I worked my way down the mountain until I became part of a small stream, which alongside my colleagues, began grew as until I was eventually part of a fast-flowing river.

Sometimes I was hurled against rocks and at other times I was spun in a whirlpool, but all the time we flowed downward and onward to the sea. (I know this because I have done journeys like it many times all over the world: the result is invariably the same).

Once we leave the mountains, we become part of a bigger river that gets deeper and wider, which flows more slowly. Looking around I see fish, vegetation, pebbles, and small water living creatures.

On occasion I look up to see boats passing overhead: This makes me realise that while I am here, I provide two of my many services; homes for fish and transportation for humans.

Suddenly a most extraordinary thing happens. I am no longer part of the river; I have been swallowed by a bird. Several minutes later I feel myself falling to earth, but not as rain or snow, I am still inside the bird!

What happened was that I had been drunk by a duck which was shot as it took off from the lake into which the river now in Italy had flowed. What a shock that must have been for the bird, but I, of course, was unharmed.

I was then picked up by a dog, which carried me to the man who shot the duck. Although the duck was dead, nothing about me had changed. I was still in the duck's stomach; the duck being dead its circulation ceased, so again I rested.

The man took the duck home where his wife cooked it. While cooking, I evaporated, thankful to have escaped having seen what happened to the duck! As the old saying goes, 'If you can't stand the heat get out of the kitchen'. (Though, I knew I could not die; I am always water - not like the poor duck.)

Once I have evaporated, I drift through an open window and return happily into Earth's atmosphere, eventually ending up in another cloud ready to embark on another adventure.

But now the wind is whisking me across the land again, carrying me higher and higher, back over the Mediterranean Sea, through the Straits of Gibraltar, into the Atlantic Ocean. On the way, while the sun passes through me, I become part of a rainbow to the delight of people on a ship below, upon which I fall, as the cloud I am in passes by, as rain.

It is a cruise ship going to Cape Town, South Africa and I have fallen onto its deck along with thousands of other raindrops.

I often wonder what shape and size I really am. I know that mostly I am round and tiny, especially when I evaporate or fall on a hard surface and splatter into minuscule pieces. I also wonder what I am: A snowflake, part of a stream, river, sea, ocean, or raindrop, or a particle within a cloud or the atmosphere?

But what about when I am none of these? When I am in human blood, juice of a fruit, sap of a tree, in the tummy of a bird, fish, or animal? Or, when I seeped deep into the Sahara Desert 10,000 years ago and seemed lost forever. Or part of the ice in Antarctica millions of years before that.

At times like these I need to be patient, an attribute I have in abundance. Not like humans for whom delay for even a few seconds can seem a catastrophe!

But my thoughts and feelings never change, even when I am in one of the great seas or oceans that cover two-thirds of the Earth, I know I am just a part of one great whole along with trillions of identical raindrops, each performing to perfection our roles in life.

This lesson of living in harmony for the common good would work for humans if they applied it.

As the ship sailed south, it got quite cold. Nothing very much happened, and I stayed where I had landed: it was too cold to evaporate. After a time, it rains again, and I get washed off the ship's surface into the ocean. Here there is a strong current that takes me down the west coast of Africa towards Antarctica; this is known as the Guinea Current. I become part of it and am being carried on a journey of some 4,000 kilometres.

As if out of the blue, we are attacked by the Benguela current coming up from the South Atlantic Ocean. The force is so strong I am pushed upwards at such a rate, in human terms, you would say I was 'out of control'!

However, I am *always under control,* working within the forces of nature, and doing what I can to enhance life.

Another lesson for humans, yet I am only a tiny raindrop!

Now I get bashed against the shores of Victoria Island, part of Lagos in Nigeria. Here I help perform an act of erosion, and because of my size, I am surprised how I can assist in changing the shape of our great planet's coastline.

How could little me do that? I am so small, surely, I hardly matter in the great scheme of things. But I know I do, and so therefore do you.

Working together in voluntary service for the overall good is an important lesson for humans.

I am equal in size to all the water particles in this gigantic act of erosion, and when we join together *and are so directed,* we become an almighty force. Almost nothing, except our *Director,* plays a greater role in life on Earth than we do.

But I am just playing my part as one of the tiniest pieces in the great jigsaw puzzle of nature's plan, known as: 'The Evolution of Life on Earth'.

I find it wonderfully humbling to know I am just the same as every other particle of H_2O. This is because raindrops don't have egos.

It is sad humans don't understand they are created equal and if they work together the results would be better than the greed and desire for power they live by. But then humans see humility as weakness and have egos.

After being buffeted against Nigeria's shores, I found myself carried out to sea by the South Equatorial current, eventually reaching the east coast of South America. I then arrived at the Uruguayan coast and headed south into the Antarctic circumpolar current.

Here I pick up speed, travelling at one kilometre an hour in the fastest of the many ocean currents. I am being driven by the Westerlies, the strong, cold winds that blow around Antarctica.

The Antarctic current circumnavigate the globe without a break. It always moves from west to east, and if I drift south far enough, I know from *bitter* experience, that I will freeze and become part of its ice cap.

This happened hundreds of millions of years ago, and I stayed frozen for four million years! If my memory serves me correctly, I recall being part of ice many times I cannot remember how many, it is too numerous.

But does that matter? Not, that I am always me and I like who and what I am.

Loving yourself is an important attribute often lacking in humans. They don't seem to understand it is impossible to love yourself when you constantly do things for selfish reasons.

Chapter 3

My Adventures Continue.

ON MY THIRD CIRCUMNAVIGATION of the Antarctic, I got swallowed by a small fish; this gets eaten by an Arctic tern about to start its annual, 36,000 km migration, from the shores of South Georgia to the Arctic Ocean: After that, it will circle the Earth on its return journey, before finding its way back to Antarctica.

During its migration, the tern dives into the sea to feed on fish, but I stayed in its blood and continued on the journey, which took approximately four months.

This particular tern flew to Greenland in the Arctic to breed where I am transferred into an egg. There I remain until it hatches several weeks later.

Soon after it hatches, I am breathed into the air by the baby bird as vapor through its beak: I notice that most of my life recently has been spent in darkness, which does not matter. Night or day, hot or cold, frozen, or unfrozen, I remain the same.

Once I reach the Arctic, I observe how it stays light all day long: although time to me is meaningless - it is always *now*.

Living in the now is important, it removes the fear of the future.

While the arctic tern lived in Antarctica it was always daylight as the sun scarcely dipped below the horizon at that time of year. So, this species of bird probably sees more daylight each year than any other creature due to spending its life exclusively in the summers of the North and South hemispheres.

How terns find their way when they are only a few weeks old, I have never understood: though I was grateful for the ride and change of scenery. I credit its *Higher Intelligence* with its incredible internal navigation, as I do with everything in my life.

If only humans knew they had the same Higher Intelligence looking after them. And how applying this would make a positive difference to their lives.

Now I am in the open, I rise once more into clouds, where the air is always cooler: this time I fall to Earth as sleet upon the Arctic Ocean. Over the next few months, the water I am in freezes and becomes part of the great ice pack of this region. The following summer, as the ice melted, I became part of an iceberg that flowed down the east coast of Greenland, eventually melting into the North Atlantic Ocean near Newfoundland.

In my time as part of the iceberg, I reflect on my past and recalled many of the things that have happened to me. Some of them extraordinary, yet, to humans, I am lifeless!

I was once a drop of water that fell from an icicle that helped form the source of the Amazon River in Peru's High Andes; not long before that I was in its thermal tributary, Shanay-timpishka, known as La Bomba, also in Peru.

That is how my life goes; at times I am frozen, at others I am boiling hot. But whatever I am, or wherever I am, it is never boring.

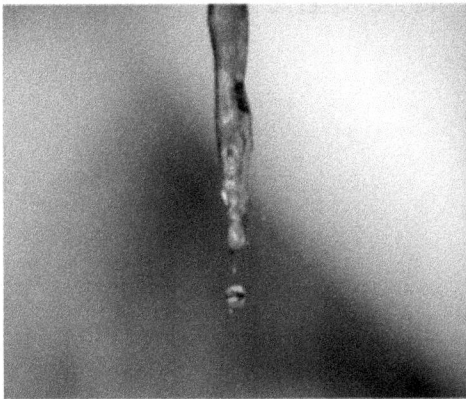

Drop of water from an icicle   La Bomba river, Peru

THE RANGE OF MY USES seems endless - not just for humans for whom I helped create electricity in the Colorado River in the USA; turned into steam in the process of making the first-ever paper in China in AD 105, and 1,676 years later, I was steam again powering James Watt's engine that propelled the UK into the Industrial Revolution.

Hydroelectric Dam Colorado River, USA

James Watt Steam Engine image

MY CONTRIBUTIONS TO flora and fauna are even greater. Not long after my experience with James Watt, in Kenya, I was drunk by a lion, in Uganda I was in a gorilla's saliva, in Greece I was eaten in an apple by a bee and ended up in honey. I was in a cow's milk on the island of Jersey, in Wales, I nourished the acorn of an oak tree, the seed of a sunflower in France. And in the Bay of Biscay, I was in the water that spouted from a whale's blowhole.

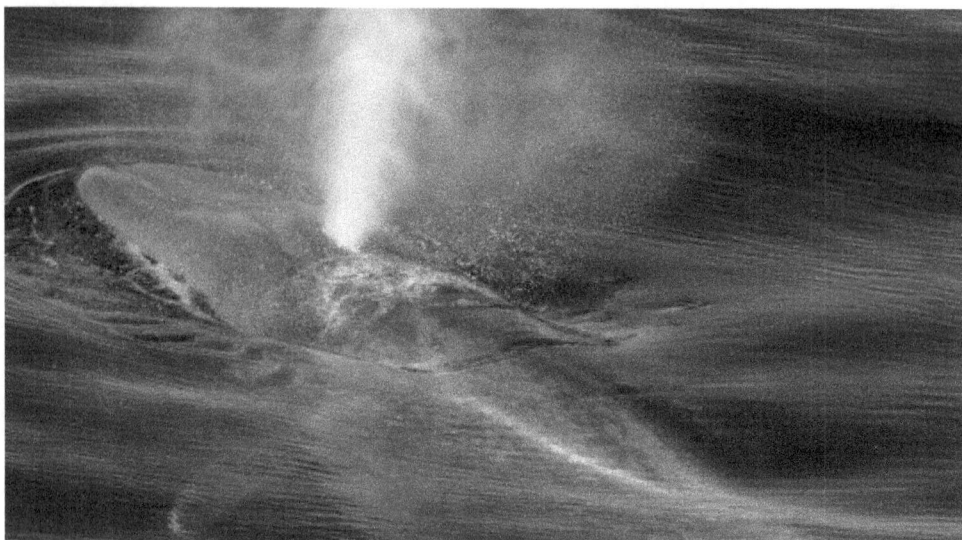

Whale spouting water from blowhole

MY JOURNEY DOESN'T stop there; I am vital in the geological world, too. I surged through a geyser in Iceland, cascaded over Victoria Falls in Zimbabwe, and long before that, over Angel Falls in Venezuela—the highest waterfall in the world. Through it all, I remained unchanged; I was, and always will be, H_2O.

And to show how versatile I am, I once flowed through the body of a dinosaur over 140 million years ago. In the 3.8 billion years I have been here, I have seen and done things beyond anything you can imagine.

Some say I landed on Earth on board a meteorite, but how I arrived here is a secret, and we droplets of water are very secretive about many aspects of our lives; just like the *Higher Power* that made and guides us.

Now, think about what comes next because it is important. Is it water's essential uses and the insightfulness of spiritual leaders which explain why the expression 'living water' is used in biblical scriptures and religious ceremonies all over the world. They seem to make clear that those who applied them had a better understanding of the spiritual features of water than humans do today.

(Remember the answer in my class to the question of dislikes: Rain was on everyone's except mine!)

Still the questions humans ask remain; *where did water come from? Who's idea was it?*

But it is not our origin that is important. What is important is to understand our freedom, ability, and willingness under all conditions to carry out our multitude of life-enhancing functions. So, maybe it's our humble strategy of not promoting our good deeds and the universe's *Intelligence* remaining anonymous that causes disbelief in *God the Creator and Living Water!*

Most people think of water as being an essential-to-life 'lifeless' liquid. But after reading my essay so far, which do you think describes it better, 'lifeless' or 'Living Water'?

If you cannot decide, let's look at what's been happening for millions of years, then prepare to face the humbling truth, as most people will laugh as they take me and my associates for granted. They see us as useful for putting out fires, quenching thirst, boiling water to make tea, coffee or food, freezing us to make ice cream, or transportation, and little else.

But note first how beneficial each of these functions is before you pursue that idea. Many people have ended up with egg on their faces doing that; and remember, we will be in those eggs too!

We are often criticised when it rains, when what we are doing is feeding the world at no financial cost. Water companies do the same but they apply the exact opposite. If only their shareholders and executives understood the truth about serving others!

The question is, do you?

On the other hand, there are water's revered marvels of natural beauty which are also free. Nothing surpasses the wonder of glistening waterfalls, the shades of turquoise, blue and white seas, crystal-clear coral reefs, enchanting tropical islands, rainbows, and the wonder of snow-covered mountains and pine trees. Plus, each of the trillions of snowflakes that fall has a different shape and design of exquisite perfection.

Iguazu Falls, South America

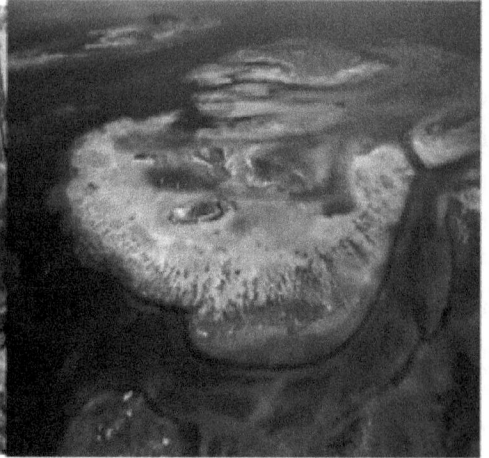

Tobago Keys Caribbean Sea

Double Rainbow   Winter Wonderland

Coral reef in crystal clear water   Snowflakes

SO, WHAT DO YOU THINK? Are drops of water really as dead as the dodo? (By the way, I existed inside the last dodo when a man killed it, and it took nearly a week for me to escape. While my brothers and sisters had the same experience with Sabre-tooth tigers and mammoths!)

Remember too before you decide, my trillions of friends and I were in the bodies of Krishna, Christ, Buddha, Mohammed, Lao-Tse, Moses, Abraham, Gandhi, Mandela, M.L. King, Mother Teresa, every saint, spiritual teacher, and all your forefathers.

We are in every member of each religion, atheist and agnostic; from Muslims and Christians to Jews, Buddhists, and non-believers we have no preferences.

So, like our *Creator,* drops of water have no human favourites; criminals, politicians, lawyers, and the mentally ill all benefit from our unconditional love: we provide life for every living thing equally. And some of us are now part of you.

Each human being, every living thing, and life-giving water, are wonders of the same Creator; so please don't knock life-enhancing raindrops. Knock instead the people, including some scientists, who don't give credit where it is due. The lessons raindrops teach about pluralism for the common good are as important to humans as they are to life on Earth.

As Mother Teresa said, 'If we have no peace, it is because we have forgotten we belong to each other.'

What she could have said is that if all are not treated equally, we have moved away from our Creator's design for humanity and life on Earth. Our roles are surely to put this right.

AS I CONTEMPLATED THESE aspects of water, most of which I had not thought of before, there was another I now realised; its constituents are eternal.

As this was the opposite of humans for whom many live on a knife edge of fear of dying, I was more in awe of water's wonders than I had ever been.

As I further considered its eternal life, I thought about oxygen, carbon dioxide, nitrogen, other natural gases, minerals, and their functions. In each instance, man hid not control their creation and distribution, yet each performs to perfection a specific role, also eternally.

Particles of water previously seemed insignificant. Now I knew each drop and its trillions of brothers and sisters were from a spiritual point of view a long way ahead of me. So, I decided to focus my attention on this until I understood it better.

At first, I wondered if knowing there are many trillions of drops of water working together that produce incredulous results means the far smaller number of humans realise that if they apply the same work ethic and do so unconditionally, mean the results would be the same. As I had experienced such success through friends in 12 Steps of Recovery Fellowships, I knew it should.

Looking at life this way makes clear that it is the self-centeredness of the roles we play that differentiates us. Thirty-eight years ago, my understanding helped me aim to eliminate the defects of character this causes that previously controlled my thoughts, motives and actions.

I thought about the role of water in my life. Not long ago, something inside me could have been in a rat, scorpion, slug, leech, worm, jellyfish, snake, wasp, mosquito, or any other creature I did not like.

Then I wondered if vegetarians realise it is the same water that they drink which was once in the bodies of crocodiles, sharks, tigers, lions, vultures, eagles, moles, anteaters, frogs, fish and birds that keep nature's flora and fauna in equilibrium and going forward? The fact is, all life on Earth is made up of a considerable amount of water, it is essential to the diet of every living thing.

Much of what I had learned were facts about water I had not acknowledged before. Maybe this is the same for you too?

Chapter 4

Raindrops on an Ocean

I RETURNED TO IMAGINING myself as a droplet of water. First is in the Sargasso Sea in the North Atlantic Ocean where baby eels, and elvers, are born. I am part of their vehicle for crossing the 6,000 miles of this mass of many trillions of other droplets of water to the rivers of Western Europe. There the elvers will grow, until at the end of their lifetimes, usually seven years, they will return to the same place in the Sargasso Sea to breed and die.

How could these elvers know from within their tiny forms the direction to travel with no parents or grandparents to guide them? I also knew they always go to the same place where their mothers and fathers had grown up. Knowing where to go provides continuous proportional distribution and ensures they go to areas with established food supplies, at the same time thereby preserving the balance of nature.

I reasoned intelligence had been involved. What was it that directed them so perfectly? Dispensing with the services of their parents at such tender ages did not seem natural either, so how could they do this? Scientists suggest their behaviour is genetically imprinted on their brains and/or in their genes: but how did that intelligence get lodged there in the first place?

I imagined their minute, elver brains, having such precisely detailed information imprinted on every gene that makes up every chromosome without any conflict because they always contain identical data. But from where did this information emanate? I was flummoxed. I had no answer.

Surely there had been elvers who made the first journey when the information was not known by their genes? It just did not seem to make sense.

It was as adults that they returned to their starting point in the Atlantic, and as hatchlings, they took their first journey!

Which had come first? I was back to the chicken and egg scenario where I had concluded after a lengthy mental debate, that the egg must have come first. But now I was confronted with a more complex problem of a baby eel travelling a massive distance, unprotected, in uncharted water, and with no installed navigational equipment. It was no wonder I was bewildered, especially as it is a migration that has happened every year for thousands of years with no breaks.

The more I tried to understand this phenomenon and everything else I had learned about evolution and spirituality, the more I came to understand the views of people who believe in animism, Gaia, and Divine Intelligence. By acknowledging the facts, the suggestion some scientists make that what has evolved was genetically transmitted no longer rang true. Creatures such as elvers intelligence had evolved, and 'something' had made that possible which could only have been achieved by strategy planning, and forward-thinking intelligence beyond anything a human mind can accomplish. Which is why it is so difficult to comprehend.

The facts surrounding migrations like that of eels, are probably part of the problem in believing in God. All too often scientists and those with high IQs are left to decide for the rest of us if an infinite invisible Intelligence exists. When they say it does or does not, we of lesser intelligence assume they may be right. But I now knew that to apply reason as to whether there is one God in charge of all there is in the universe was wrong. Such knowledge is only realised by those who have a spiritual awakening and have practised prayer and meditation for years. As this applies to anyone who practices these principles, it is these men and women who should be listened to for such guidance.

If you think my conclusion is wrong, try answering this question. How could intelligence-gathering cells in the elvers have formed in the first instance without there having been *Intelligence* behind this feature of their anatomy?

It was by looking at it this way, that I realised intelligence, though invisible, was an entity that went backwards into infinity. Elvers' intelligence did not appear by magic unless there was a magician to have produced it! Then I thought, '*God could be a magician too, if He, She, or It, chooses to be.*'

As I pondered on the idea of *God* performing '*magic*,' I was drawn to think about plants. I grew up watching dandelion seeds be carried effortlessly by a breeze once they had completed their life as a flower. Without seemingly knowing where they were going, they spread almost everywhere. And although often thought of as weeds, dandelion leaves, roots and flowers are nutritious and have health benefits, ranging from healing inflammation to preventing liver damage.

I wondered, *'Were these accidental, or some of God's magic tricks?'*

Dandelion meadow

Dandelion seed parachutes

Dandelion leaves & roots

Dandelion milky stem sap

BUT DANDELIONS ARE not the only plant with human health benefits, globally there are many with thousands of known uses.

In forests in Southeast Asia and South America there are approximately 8,000 plants used to help fight diseases; these range from cancer treatment, soothing asthma, reducing anxiety to treating gallstones.

As the medicinal uses from trees and plants where water is a key ingredient, includes fruit, cannabis, poppies and cocao, given the facts, I found some natural wonders especially interesting. The *God* I had come to believe in had imagination and a sense of humour!

My first thought turned to alcohol. I thought of grapes, apples, pears, hops, barley, elderberries, and other fruits and vegetables used to make the liquid from which I nearly died overindulging in. Yet, I also knew alcohol is used medicinally as an antiseptic, pick-me-up, disinfectant and antidote. It is only people like me with an allergy to it for whom it was dangerous.

Grapes for wine

Cider apples

ON THE OTHER HAND, many use cannabis as a treatment for chronic pain, nausea caused by chemotherapy, muscle spasms caused by multiple sclerosis, severe forms of epilepsy and mental health conditions. Scientists in the beauty industry have found that its components have many anti-aging properties that work well in hand creams and hair care products. In addition to the benefits of cannabis in skincare, it has significant hydrating properties. While features of its flowers are used to help people relax, as a sedative, and as an anti-depressant when inhaled.

Cannabis skin and body products

Cannabis as medicine

POPPY SEEDS ARE RICH in healthy plant compounds and nutrients like manganese. These seeds and their oil may boost fertility and aid digestion. The opium poppy is an annual medicinal herb. It contains many alkaloids that are frequently used as an analgesic, anti-tussive and anti-spasmodic in modern medicine. Besides, it is also grown as a source of edible seed and seed oil.

Opium poppies

CACAO BEAN POWDER IS packed with flavonoids, which have been shown
to help lower blood pressure, while improving blood flow to our brains and heart.
They also help to support gut bacteria, may decrease the chance of heart disease
and improve blood-sugar balance, as well as improving the condition of skin.
They are rich in magnesium which is good for bones and teeth.

Cocao pods with beans inside

FROM MY PERSPECTIVE as a water droplet, to understand this better, I imagined I was a sunflower seed in France and had settled several centimetres below the surface of the ground the previous year. It is now spring, and I have been lying there dormant ever since. Much of the time I felt as dry as a bone, but earlier today it rained, and now I feel a dampness surrounding me. (The raindrops that fell, seeped through the soil until they wrapped themselves around me.) This is causing *something* inside me to stir.

The next thing I know, I am feeding from the water and other invisible life-giving properties outside my shell, and from these I begin to grow. Unaided by me, I have unwittingly absorbed whatever these properties were.

It leads me to a question.

How did my sunflower family's first seeds know there were life-giving properties in soil when they fell to the ground centuries ago? In addition, that they were essential to meet their needs!

It is now mid-summer and I have grown to a metre and a half in height. Since I germinated, I have fed on the invisible ingredients in rain, soil, and air, and helped to grow by the heat and light from the sun. As I wondered how had I been able to do this, I realised it was not *me*. I had been blessed this way; the earth, air, water, heat and light upon which I thrived were, *manna from Heaven*.

The big question is, would these minerals and other forces I needed to go from a tiny seed to a flower, have required *Intelligence* to be sure they were in the right place at the right time to perform their functions?

The only answer I could think of was, 'Yes, I knew I didn't do it. But *something* had made it happen.'

Now, I wondered, 'Would I need intelligence to create my knowledge of the existence of these life-giving properties?' Again, the only answer I believed was, 'Yes.' And, lastly, 'Could it be just a string of many coincidences that caused all my needs to be perfectly positioned for my life to have evolved and then fulfilled? After all, I am loved for my uses. I am in many brands of vegetable oils, spreads and loved by finches.'

'No, that would be impossible.'

Sunflowers

Goldfinch eating seeds

Sunflower seeds

Chapter 5

Water's Uses

IN MY COMPANION BOOK *The God Gap,* I wrote a chapter when I imagined I was a drop of water and described some of my life: that was twenty-five years ago. Since then, that drop of water would have been involved in many geological and ecological issues. As water's role goes largely unnoticed, I have written a biographical chapter of what some of them might have been.

'What a difference a day makes, while a week in my life as a drop of water can be truly remarkable. For instance, on September 9th, 2001, I was one of billions of water droplets used to extinguish the fire at the Twin Towers in New York. A few hours later, I found myself in the Hudson River, on my way to join the northbound Gulf Stream in the Atlantic Ocean. Once I joined it, I moved up America's east coast to Canada, where, in the Gulf of St. Lawrence, I was swallowed in a massive mouthful of water by a whale.

Can you imagine anything more diverse happening to a human? Yet for me, such experiences are just part of everyday life. I know my functions are always crucial, but to humans, they mostly go unnoticed and are usually taken for granted. What they need to figure out is this: how is it that I am always in the right place at the right time, and in no instance are they responsible for what I truly accomplish? So, who or what is responsible? To answer that, understanding the water cycle should help.

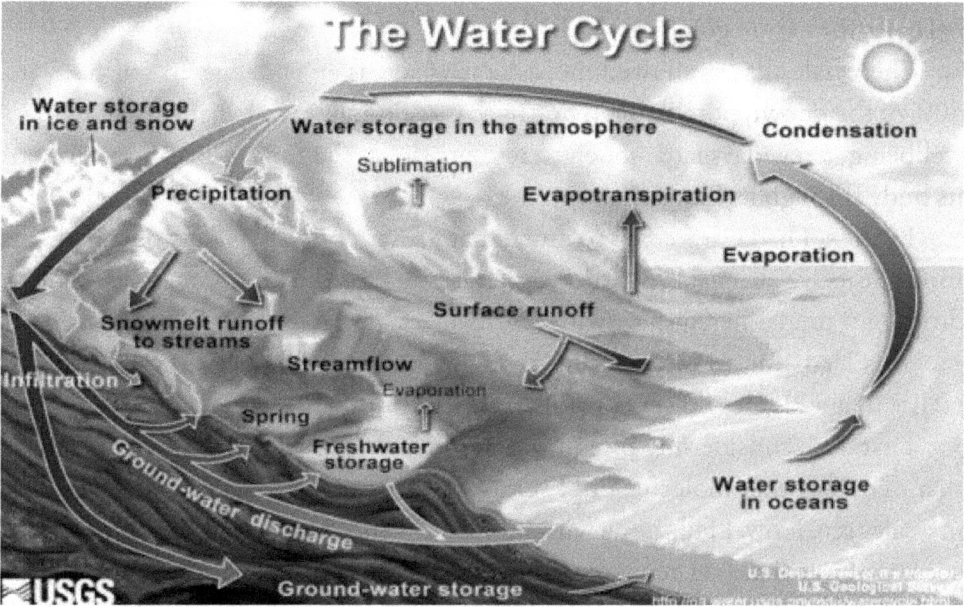

The Water Cycle

Water storage in ice and snow

Water storage in the atmosphere

Condensation

Sublimation

Precipitation

Evapotranspiration

Evaporation

Snowmelt runoff to streams

Surface runoff

Infiltration

Streamflow

Evaporation

Spring

Freshwater storage

Ground-water discharge

Water storage in oceans

Ground-water storage

USGS

U.S. Department of the Interior
U.S. Geological Survey

Water Cycle

LET'S BEGIN IN THE oceans and seas, where I serve many roles for life, including maintaining precise salt concentrations. You might wonder how this balance came to be, but there's an even bigger question to answer: life on land depends on pure, natural water. How was that to be achieved when two thirds of the world was covered with salt water?

To do this, the perfect mechanisms are in place: heat, at just the right temperatures causes the perfect amount of water to evaporate, rise, and condense into clouds. Wind and gravity then take care of the rest.

Is all of this just chance, you should ask? The answer, of course, is no, though the magnitude of it does defy human logic.

Another thing to note is that life on Earth only exists on a habitable surface; whereas, life in the oceans, seas, and lakes goes to great depths, and all of this is only possible thanks to me and my trillions upon trillions of siblings. I trust that the next time we are falling on you or a strong wind is blowing you will refrain from criticism, what us and the other elements are doing is providing the right balances for nature going forward.

The whale that swallowed me, a right whale, is an endangered species, with 40% of the known population living in that region. But while I am essential to its wellbeing, a few days later, much to my surprise at the time, it blasted me out of its body through its blowhole!

By then, I had warmed up so much that I rose, along with many of the whale's former captive droplets. A little while later, I joined a cloud, and many more drops of water gathered there with me.

This leads me to a staggering thought, one that I hope will help humans realise that a Higher Power controls everything.

First, let's consider my role in the water cycle, which, overall, is the same as that of every other molecule of H2O. A few days ago, I was a crucial part of life in a salty sea. Now, I fall, along with billions of other drops of H2O, on the south coast of Greenland. (I recognize this place because the last time I was here, I froze and stayed frozen for months—always wondering why it's called Greenland and not Whiteland! That's humans for you. No wonder they get so many things wrong!)

This time, however, it was very different. I was absorbed by a tuft of grass, and soon, a white-fronted goose came along and "ate" me. The next thing I knew, the goose was joined by two others of its species, and a day later, we began what I now know was a migratory journey to Slimbridge Wetlands on the Severn Estuary, on the border between England and Wales. This was a good choice for the geese, as Slimbridge is an outstanding nature reserve, the former brainchild of the conservationist and exceptional wildfowl artist, Sir Peter Scott. Here, these rare geese to the UK will spend the winter, knowing that all their needs will be met.

Greenland White-Fronted Geese

Slimbridge Wetlands

AS THE SAYING GOES, "time flies." We traveled 3,000 kilometers at 70 kilometers per hour in just a few days without a single stop. Then, suddenly, like being in an airplane, I realised we were descending. The next thing I heard was the "woosh" of the geese's webbed feet hitting the water. After a few seconds of skimming across the surface, we came to an abrupt stop.

For most of this journey, I had remained in the goose's stomach as it digested the grass. Now, I was passed into the water in its droppings, where I felt at home again. After a while, I decided to explore; I enjoy traveling, but I also like to know where I am, as sometimes I am not so keen on my surroundings! Though I always know I am precisely where I'm meant to be, nurturing nature's needs.

It was my first time at Slimbridge Wetlands. What I hadn't realised was that it was like a gigantic airport for long-distance travellers, heading in all directions—north, east, west, and south. In my case, my next stop was Tangier, Morocco, in North Africa.

What an experience this turned out to be. I had been ingested by a stork at Slimbridge that nested on the rooftops of this Atlantic coast city. There, I was laid in an egg. After hatching, I could see Spain to the north, while I knew from past experience that the Sahara Desert lay to the south and the Rif Mountains were to the east. I knew it was unlikely I would be blown south - I have not often fallen on deserts - and hoped that when I evaporated, I would be carried toward the Rif, one of my favourite places.

The reason I am so drawn to the Rif Mountains is not just because of their stunning beauty, but because they are home to some of Africa's most impoverished people. If I can do anything to help them—something far more than Morocco's king has done—then I will have fulfilled one of my life's purposes. If only people could understand that helping those less fortunate benefits the giver just as much as the receiver, the world would be a far better place. And if it takes a drop of water to make that clear, then only God can reach those who remain so hopelessly unenlightened.

What happened next is something that has stayed with me more profoundly than any other experience in my life. I will never forget the time, the place, or the day. Though strange things often happen to me, from a human perspective, this was the most tragic.

Earlier that day, I had been in a glass of water with a poor fisherman, Mouchcine Fikri, in the town of El Hoceima on Morocco's Mediterranean coast. Mouchcine had just returned from his catch—swordfish—which was his family's only source of sustenance. But his hopes were shattered when police seized his catch and dumped it into a passing garbage truck.

In a desperate attempt to salvage his livelihood, Mouchcine leaped into the back of the truck to recover his fish. But tragically, he was mangled to death in the process. The inhumanity of the way he died caused widespread unrest in Morocco and brought home to many a side of Morocco and its playboy kings' lifestyle hitherto unknown. Just selling some of Mohamed V1th's palaces, $ million watches and fleet of luxury cars would solve much of this country's poverty problem!

In mid-September 2023, a shower of rain fell near the banks of the Jordan River in the Holy Land, and I was part of one of those raindrops. Over the following days, I flowed along the western bank of the river, eventually reaching the Palestinian side of the Dead Sea, near the site where, in the late 1940s, a young Bedouin shepherd made what is often considered the greatest archaeological discovery of the 20th century - the Dead Sea Scrolls.

However, this is not the only reason why this area is one of the most significant in human history.

While the Jordan River may not be as vast or as famous as the Amazon, Nile, Ganges, or Mississippi, which incidentally I had visited before, its spiritual importance is unparalleled. The river's width averages about one hundred feet, and its depth ranges from three to ten feet. Yet, it is truly a *living river* for at least two reasons.

JORDAN RIVER, DEAD Sea Palestine, Israel & Jordan

FOR CHRISTIANS, THE Jordan's waters are considered life-giving because it was here that Jesus was baptized by John the Baptist. Along my journey, I observed sixteen species of fish, each unique to this remarkable river that nourishes the land in its valley as it flows south toward the Sea of Galilee. So, I already knew that calling it a "living river" was quite fitting!

At the southern end of the Sea of Galilee, the Jordan River begins its main course - before it reaches the Sea of Galilee, it is made up of streams. Afterward, it meanders lazily, winding a path of 223 miles through a valley made fertile by its waters, before reaching the Dead Sea.

For most of its length, the Jordan River forms the eastern boundary of Israel. In the days of Moses, it was at the Jordan where the Israelites crossed into the Promised Land. It remains a boundary today: the east bank is Jordan, and the west bank is Israel. Much of the current conflict centres around the southern region, where part of the West Bank is inhabited by Palestinians.

It is here that the Dead Sea salts and soaps are made, renowned around the world for being some of the best skin care products available. And what do they need to perform their functions? They need pure, natural water. Do you think of this when you use them? I would be surprised if you do. But if you do, consider: who is the *Real Provider?*

Additionally, Palestinian olive oil is considered some of the finest in the world, and this is only made possible because of the water flowing through the Jordan Valley.

But I, and my fellow water droplets, serve other purposes as well. So, let us explore some of these functions further.

The buckling of the earth's crust, combined with volcanic activity and erosion, formed mountain ranges over millions of years. As warm air rises, temperatures drop, causing the water vapor it contains to condense into droplets on particles of dust and pollen, forming clouds. If the air cools enough, these droplets turn into ice, eventually falling as snow.

Long after these natural processes shaped the earth, humans arrived. When they saw the mountains and snow, they discovered by the end of the 19th century that attaching narrow strips of wood to their feet would propel them down snowy slopes more efficiently than walking, thus skiing was invented. Today, an entire industry has evolved around skiing, providing work and pleasure for millions worldwide. This industry has even led to the creation of towns and villages. Yet the benefits and significance of skiing for humanity extend further still.

Ski resorts require extensive networks - roads, railways, air transport, and ski lifts - alongside specialised clothing, equipment, and accessories, all of which have arisen in barely a century. Once barren, often inaccessible mountains are now centers of human activity and enjoyment.

Wooden skis   Modern ski-ing.

ALTHOUGH MOUNTAINS, snow, and skiing developed independently, the relationship between them can seem as if a *Higher Power* foresaw platforms and resources that would later serve humans.

Just as humans discovered skiing, they also observed how wood floated on water. Although they did not create wood, they saw its potential, crafting boats that met needs for fishing and transportation. Wooden boats propelled by oars and sails allowed humans to reach distant waters for fishing, and larger boats began carrying passengers and goods. Over time, adaptations to these primitive crafts led to metal and fiberglass ships, made possible by materials available in abundance.

As humans evolved, they discovered other energy sources like oil, coal, gas, water, and electricity to power larger vessels, made from plentiful materials suited to their needs. Studies of fishing boats reveal adaptations suited to local conditions, affirming that human intelligence around the world contributed to this advancement. This suggests that a *Higher Power* provided resources globally, allowing people across different regions to develop similar solutions using local materials - examples of early boats from Europe, Asia, and Africa still demonstrate this today.

Coracle in Wales

China

Africa

Kerala, India

WITNESSING BIRDS, BATS, and insects defy gravity, humans wondered if they, too, could fly. Through their ingenuity, they realised flight could unlock new possibilities, leading the Wright Brothers to invent the airplane. This invention eventually led to rockets and satellites with telescopes to explore space. Unlike animals that fly, human-made flying machines required materials and designs humans didn't initially conceive of - pointing to a *Higher Power* that provided the resources long before humans understood them. Humanity used its intelligence to harness these resources, achieving remarkable results, but only after a foresighted *Intelligence* had made those resources available.

This realization that God's presence is continuous and purposeful has deepened my understanding. God's aims as a loving Creator became a comforting reality for me, reminding us of life's larger purpose.

A striking example of evolution in action is watching trees in autumn change colour, then drop their leaves, which decay into nutrient-rich soil, helping sustain life. While the colour transformations are subtle, they are beautiful, delighting observers. However, as human activity reduces the world's deciduous forests, such scenes may become rare. Additionally, without decaying forests to renew carbon deposits, future generations may lack current energy sources, emphasising the need for sustainable alternatives.

Forests are vital - they close the emissions gap, harbour biodiversity, regulate climate, and contribute to Indigenous cultural, economic, and social life. The *Higher Power* is aware of this, and while humans are accountable for environmental changes, we must also take action to restore balance.

In Earth's 3.5-billion-year history, there have been vast transformations that led to the biodiversity and beauty we see today. This gives hope for the next 3 billion years, though dramatic changes in nature can occur quickly. Floods, fires, avalanches, storms, and volcanoes remind us of how unpredictable and powerful nature is, particularly over millennia.

Human progress in the past century has allowed for breakthroughs like electricity, telecommunication, and transportation, all possible because the materials needed existed first. This realisation shows that everything on our planet has a purpose, prepared long before humans arrived. Fossil fuels, now vital to modern life, derive from ancient decaying plants that grew in precise locations, aware of future needs. Many natural phenomena - from mountains to rivers and weather patterns - emphasise that life is sustained by resources far beyond human control.

While it may appear that no new minerals are forming, untapped reserves remain available for humanity's future needs. Science tells us Earth has vast untapped resources, especially in regions like the Middle East's Red Sea and North Africa's High Atlas, where petroleum, metals, and minerals are abundant and may hold new uses. And developing medicines from the cannabis plant is high on many country's agendas.

Earth's *Higher Power* has consistently provided sustenance for flora and fauna and has orchestrated physical and meteorological conditions with precision. As time progresses, humans will find new food sources and health remedies, guided by the same *Higher Power* that has already provided for human needs. In this light, God's impartial provision for all humanity urges equitable sharing, affirming that love is the foundation of life. If some are blessed with more, they should work to ensure food, water, and medicine are accessible to all - an essential humanitarian principle. Unnecessary famine, especially when weaponised, is abhorrent both to human dignity and to a loving Creator.

Given God's loving nature, we as humans must strive to reflect that love. While modern man is the most advanced being to inhabit Earth, we must learn from history's mistakes and realise that unchecked greed and power could be our undoing, much like several past regimes. Progress in the Western world and improvements in developing countries offer some hope, though it is clear we still have work to do.

The *Higher Power* continues to support human intelligence, allowing us to harness resources and work toward a balanced world. Still, eradicating systemic problems may require Divine Intervention. Humanity's role is to advance what is good and rectify what is harmful, aligning our actions with the *Higher Power's* vision for Earth. Only by doing so can we fulfil our potential as the stewards of this diverse, abundant world.

I then compared this to how water has been used to shape and sculpt the Earth's landscape. Rivers, tides, streams, rain, hail, snow, ice, storms, and waterfalls are some of the many ways the *Creator* has employed water to carve awe-inspiring geographical features. These breathtaking formations, born from the interplay of water, time, and the environment, not only serve a purpose but also bring joy to humanity - and, I imagine, to their Creator as well.

From the majestic Victoria Falls in Africa to the thundering Niagara Falls in North America, from the pristine white travertine terraces of Pamukkale in Turkey to the iconic basalt columns of the Giant's Causeway in Ireland, and from the dramatic Verdon Gorge in France to the vast expanse of the Grand Canyon in the USA, the wonders of water's handiwork are celebrated worldwide. Consider also the ethereal ice caves scattered across Alaska, Argentina, Austria, China, Iceland, Japan, New Zealand, Russia, Slovakia, Ukraine, and beyond - each a testament to nature's extraordinary artistry.

Mineral Rich Thermal pools Pamukkale Turkey

Verdun Gorge France

Vatnajokull National Park, Skaftafell, Iceland

THE REALISATION OF God's global, continuous, and purposeful presence has profoundly deepened my understanding of life. Recognising the Creator's intentions has become a comforting reality, reminding me of a greater purpose: to love all that exists.

This truth resonated with me deeply today, January 8, 2025, as well as during my recent time in the USA. While fires raged in Los Angeles, I was part of the water cannons that, along with trillions of my kindred droplets, helped extinguish the devastating flames. Just weeks earlier, on Thanksgiving Day, I found myself in the Chardonnay poured at a Californian home, bringing joy to the celebration.

If these moments aren't reason enough to appreciate water, consider this: I may have been part of the amniotic fluid that cradled and protected you in your mother's womb before you were born. There, I played an essential role in your embryonic development. Simply put, without water—this living, life-giving essence—you would not be here today.

Firefighter battles Pacific Palisades fire L.A. January 2025.

Californian Chardonnay

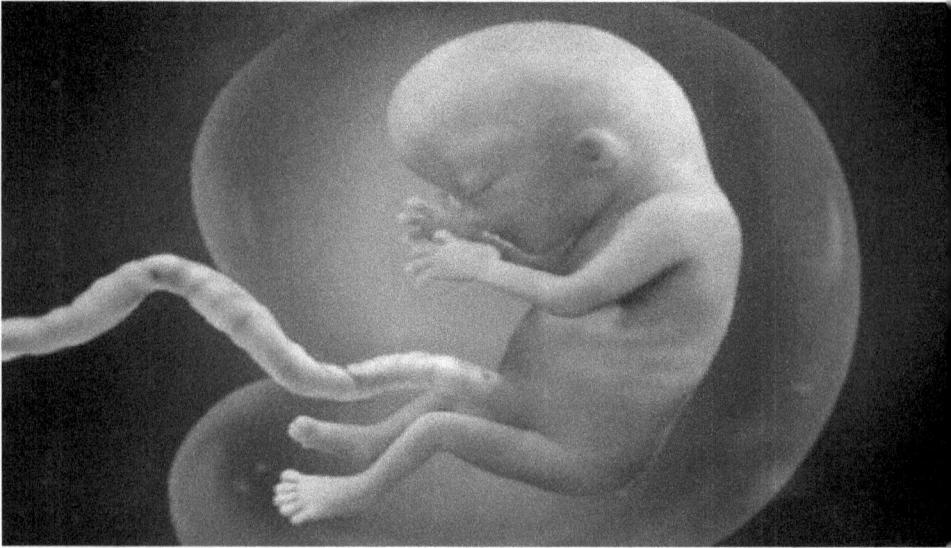

HUMAN EMBRYO SURROUNDED by fluid in its mother's womb.

Afterword.

There has always been enough of everything to meet Earth's needs. Moreover, the needs of all animals, birds, fish, reptiles, insects, and plants are continuously met. How is it that, at all times, there exists the perfect balance of oxygen, hydrogen, carbon dioxide, water, gas, oil, coal, stone, slate, wood, lead, granite, magnesium, bronze, flint, copper, rubber, salt, mercury, gold, platinum, silver, cotton, leather, wool, and all forms of food and nutrients? This includes vitamins, minerals, protein, and fibre - all essential for Earth and its inhabitants to thrive. Yet, as Mahatma Gandhi observed, while the Earth provides enough for everyone's needs, man's greed has disrupted the fair distribution of these resources.

I found myself reflecting on the vastness of the solar system - its sizes, speeds, and distances - and comparing this to the intricate wonders of nature on a microscopic scale. Despite scientific advancements, the origins of most of these marvels remain unexplained. Science tells us that everything in the universe originated from gas and dust, which only deepens the mystery and makes it harder to dismiss the possibility of a guiding intelligence - what some might call a Magician.

The more I contemplated the universe and Earth's intricate systems, the more I was moved to marvel at them. No longer content with mere observation, I began to delve into their spirit, exploring dimensions I had previously ignored. As I opened my mind to the possibility of a *Creator,* I discovered that the more receptive I became, the more wonders I witnessed. It felt as though my prior closed-mindedness had kept the door to this realm firmly shut. This realisation inspired me to seek deeper understanding - to see if I could truly find evidence that *God* exists, always has, and always will.

To begin, I studied world religions, exploring their teachings and their connection to human evolution, hoping to uncover answers. I bought *The World Religions: A Lion Handbook*, which provided a concise overview of each faith. Having been raised in a Christian environment and being married to a Jewish woman, I already had some familiarity with these two religions.

Two years later, while staying at a bed and breakfast in Herefordshire, I met a couple, Paul and Fatima, who had converted from Christianity to Islam. Paul and I began a regular correspondence, and after purchasing an English translation of the Qur'an, I gained insights into the teachings of Muhammad and found beauty in Islamic principles.

A few years later, I met Joanne, who was attending Buddhism classes in Cheltenham. I accompanied her to several sessions and, through additional reading, gained an appreciation for the philosophical teachings of Buddha. I was struck by how similar they were to the wisdom found in Christianity, Judaism, and Islam.

Hinduism, however, proved more complex. I attempted to read the *Bhagavad Gita* and visited India three times to better understand its intricate teachings. Over time, I realised that its core principles aligned closely with those of the other religions I had studied.

Eventually, I concluded that all religions are equal in their essence and offer profound guidance for humanity. Each contains exemplary tools for living and wisdom that enriches our understanding of existence. After discovering a faith that resonated with and unified these doctrines, I began attending Quaker meetings and, after several years, embraced their way of life.

As I reflected on all this - the perfection of the universe and the profound teachings of these faiths - I began to wonder if the *Higher Power* behind creation, which I call *God,* operates through a host of angels. Perhaps these angels are entrusted with overseeing the countless functions of creation, each embodying God's will to maintain the balance and beauty we witness in the world. In doing so, they may also help ensure that love continues to grow and reach those who remain unenlightened.

1. Lessons of Living Water for Humans.

Water is essential to life on Earth and has profoundly shaped human society through its properties and cycles. These 12 lessons from water offer timeless wisdom that, if applied, could benefit individuals, families, and humanity as a whole.

1. Harmony
   Raindrops are part of a seamless, balanced cycle - no competition, no waste. This reminds humans of the importance of living in tune with nature and fostering cooperative relationships.

2. Patience
   Water doesn't act hastily. It carves mountains and nurtures seeds over time, showing us the value of trusting the process and embracing the pace of natural growth.

3. Timing
   Water exists fully in each phase- vapor, liquid, or ice. It embodies mindfulness, teaching us to focus on the present rather than worrying about the past or future.

4. Service
   Water nourishes life selflessly, expecting nothing in return. This mirrors the human ideal of altruism - serving others for the greater good.

5. Selflove
   Every raindrop has a unique role, just as every person contributes uniquely to the world. Embracing self-worth is key to personal and collective fulfillment.

6. Togetherness
   Raindrops form rivers and nourish ecosystems collectively. This illustrates the power of teamwork and community in achieving greatness.

7. Connection
   A single drop of water evaporates quickly, but together, they endure and thrive. This highlights the human need for relationships and

community.

8. Humility

   Water serves silently and ceaselessly, transforming landscapes and sustaining life without ego. Its quiet resilience shows that true strength lies in humility.

9. Egoism

   Raindrops don't claim individual credit; they merge into the whole to fulfill their purpose. This is a reminder for humans to balance individuality with collective good and temper ego with humility.

10. Balance

    Water exemplifies perfect balance, sustaining all life in every ocean, sea, and stream with precise provision. This challenges us to seek balance in our own lives and consider the possibility of higher intelligence behind nature's design.

11. Perseverance

    Water never gives up, no matter the obstacle. Like water, we should persist, embodying the adage: "If you fall seven times, get up eight."

12. Anonymity

    Each raindrop performs its function without seeking recognition, knowing its value is equal to all others. This encourages humans to embrace humility and focus on purpose rather than seeking accolades.

Did you love *Raindrops From Heaven*? Then you should read *Natural Theology - Exploring Nature through a Spiritual Lens* by Robert Tennant-Ralphs!

Natural Theology - Exploring Nature through a Spiritual Lens is the journey of an ordinary man who led two extraordinary lives. By merging facts about natural phenomena and spiritual practices, he bridged the gap between scientific theory and the reality of nature. He came to believe in a Higher Power behind all that exists.

   In this accesible book, he shares the path he took and argues that anyone can achieve similar revelations – whether by chance as he did, or through intentional practice.

Also by Robert Tennant-Ralphs

The God Gap
The God Gap
The God Gap
Natural Theology - Exploring Nature through a Spiritual Lens
Natural Theology: Exploring Nature through a Spiritual Lens
Raindrops From Heaven

www.ingramcontent.com/pod-product-compliance
Lightning Source LLC
Chambersburg PA
CBHW060049050426
42448CB00011B/2359